Monitoring of Power System Quality

Project done by the Student
Bongani Dlamini
Electrical and Electronic Engineering Department
University of Swaziland

The project examined and corrected by
Dr Hidaia Mahmood Alassouli

CONTENTS

ABSTRACT

Power Quality can be defined as the characteristics of the electricity at a given point on an electrical system, evaluated against a set of reference technical parameters. These parameters might relate to the compatibility between electricity supplied on a network and the loads connected to that network.

The voltage waveform is normally distorted, and we have the so called Power Quality disturbances such as; voltage dips/swells, transients, harmonics and voltage unbalance amongst others. The study of Power Quality encompasses the Power Quality disturbances, as well as Power Quality standards, and Power Quality Monitoring.

This project will tackle the subject of Power Quality, Power Quality Disturbances, Power Quality Standards as well as Power Quality Monitoring. A general description of each of the disturbances will be given, and the basic techniques which are used to mitigate that disturbance so as to improve the quality of the supply are presented.

CHAPTER 1 INTRODUCTION

1.1 Introducing the research title

Power utilities in every country on the globe have a principal mandate of supplying electrical power to a diverse range of customers. By diverse we mean customers who demand power on varying standards; residential, commercial, as well as industrial. Residential customers normally have a different set of demands in as far as power supplied is concerned, as compared to industry. The common thing on these types of customers is that they call for a disturbance-free supply of electrical power.

Industrial customers can experience interruptions to important processes during momentary sags associated with remote faults on the utility system. Commercial customers are installing high efficiency and electronic office equipment resulting in higher harmonic levels in the buildings. These harmonic sources cause excessive neutral currents and transformer overheating. Even residential customers are concerned about surge protection for sensitive electronics in the house and the impact of momentary interruptions on their electronic equipment.

Noted on the previous paragraph are some of the factors that can disturb the power supplied to the customer. These factors are sags and harmonics. Factors that can cause disturbances to the supply of power are not limited to aforementioned two, but there are a few of them which need attention from power utilities.

As the utility industry undergoes restructuring and all customers find their service needs changing with increased use of equipment and processes more susceptible to power system disturbances, power suppliers and customers will find a solid background in Power Quality (Power Quality) not only useful, but also for continued productivity and competitiveness.

The subject of Power Quality therefore becomes a matter of utmost importance to every power utility that needs to supply an up to standard electrical power to its customers. Power Quality by definition is a set of electrical boundaries that allows equipment to function in its intended manner without significant loss of performance or life expectancy. Perfect Power Quality is characterized by a perfect sinusoidal voltage source without waveform distortion, variation in amplitude or variation in frequency. It is then mandatory for the utility to provide a perfectly sinusoidal voltage source, with constant frequency and amplitude.

There are however factors that can cause deviations from this perfectly sinusoidal voltage source and these are the factors that every power utility should be wary of from time to time. These factors distort the voltage waveform by changing the amplitude or cause deviation in the frequency. To fully comprehend the subject of Power Quality, these factors must be studied and monitored thoroughly, and measures to mitigate them be put in place.

The factors that are a threat to Power Quality are called Power Quality disturbances and are amongst others; voltage dips, voltage flicker, harmonics, frequency variations, voltage swells, over-voltages, under-voltages and transients. A system free from these Power Quality disturbances is what every power supplying utility should strive for in order to have a good quality of supply to its customers.

Swaziland Electricity Company (SEC) being the provider of electrical power in Swaziland is as well striving to supply a distortion-less voltage source to its customers, so she needs to give the subject of Power Quality some utmost attention.

1.2 Problem Statement
In her principal impetus of providing a distortion-less supply of electrical power to its customers, SEC needs to make sure that the Power Quality disturbances are minimized on the network. So a study on Power Quality can go a long way into solving existing Power Quality disturbances as well as problems that may be experienced in the future. The company also needs to do some gap analysis on Power Quality. This gap analysis will help SEC to gauge her stand on Power Quality now, set some goals and find measures that need to be taken to attain those goals.

1.3 Research Objectives
This research has three major objectives:

➤   Study the subject of power quality and focus on power quality disturbances; their effects on the network, as well the basic techniques used to mitigate them and improve the quality of supply.

➤   Monitor the power quality disturbances on SEC's network.

➤   Compare what literature states on Power Quality with what actually happens on the network.

After the research I intend to come up with some recommendations that if implemented will go a long way in solving SEC's Power Quality problems now and in the future.

1.4 Hypothesis

An in-depth analysis of data collected from monitoring the network for power quality is vital in making decisions for preventive and corrective maintenance on the network.

1.5 Methodology

The following methodology was carried out for this project:

1. Carried out a literature review on the subject of Power Quality, with emphasis on power utilities. The literature review made it possible to be abreast with Power Quality issues from the electrical distributor. This included studying international and local Power Quality standards documents.

2. Gathered some data from the SEC's transmission network for a period of three months from six substations located on the 400kV/132kV/66kV network.

3. For the purposes of this report data on Voltage Dips, Voltage Unbalance and Harmonics was of main interest, due to time constraints.

The data analyzed tor this research covers the period from January 2014 to March 2014. Data from this period is data enough to make some informed conclusions on the status of the network, though only daily and weekly variations in Power Quality levels can be identified. With Power Quality being continuously monitored by SEC, there is a great potential for on-going studies to broaden the scope to identifying monthly, seasonal and annual trends in Power Quality levels.

1.6 Thesis Structure

Chapter 1 was the introductory chapter where the research topic was introduced and further unpacked. The motivation to carry out this research in the form of a problem statement is presented. The significance of the study is shown in the research objectives and hypothesis subsections. Then lastly we present the mode of operation that made it possible to write this paper in the methodology section.

In Chapter 2 current literature is reviewed so as to be informed on the three Power Quality

7

parameters of interest on this paper. Each parameter is studied under the sub-headings; definition, causes, effects and mitigating measures.

Power Quality standards are discussed in Chapter 3, local and international standards formulated by accredited world organizations like the IEEE. The measurement and analysis of Power Quality levels is incomplete without reference to specified limits. Established international and local standards exist for the measurement and quantifying of Power Quality data.

The instrumentation requirements for monitoring Power Quality are discussed in Chapter 4. The issues of what should be measured, where and for how long are discussed. The requirements of the instruments used in the Power Quality monitoring are described with reference to relevant standards. Details of the monitoring program at SEC are presented.

Chapter 5 includes the data collected being presented, discussed and analyzed.

The closing chapter includes the overall conclusions of the study, where recommendations are made. Suggestions for future work that is capable of building on this study are included in Chapter 6.

CHAPTER 2 LITERATURE REVIEW ON POWER QUALITY DISTURBANCES

2.1 Introduction

There can be a number of reasons why we need to study power quality, either as power suppliers, or as power consumers. Some of the many a reason is listed below as an introduction to my literature review. These are:

> - Modern electric appliances are equipped with power electronics utilizing microprocessors/microcontrollers. These appliances introduce various types of power quality disturbances and moreover, the very same appliances are very sensitive to the power quality disturbances.

> - Industrial equipment such as high efficiency adjustable speed motor drives and shunt capacitors are now used extensively. The complexity of industrial processes, which result in huge economic losses if equipment fails or malfunctions.

> - Complex interconnections of systems, resulting in more severe consequences if any one component fails. Moreover, various sophisticated power electronics equipment which is very sensitive to the power quality disturbances, is used for improving system stability, operation and efficiency.

> - Significant increase in embedded generation and renewable energy sources which create new power quality disturbances; voltage variations, flicker and waveform distortions [2], [3], [4].

This chapter looks at the various categories of power quality disturbances categorized as Transients, Long Duration Voltage Variations, Short Duration Voltage Variations, and Waveform Distortions. Stability as a factor to consider in power quality studies is also defined in this chapter. Basic methods to mitigate power quality disturbances, as well as improving the quality of supply are defined. Additional information is considered for voltage dips, harmonics and voltage unbalance.

2.2 Power Quality Disturbances

2 2.1.Transients

These are any undesirable events that are momentary in nature. They can be defined as either impulsive or oscillatory.

- ➤ Impulsive Transients: these are sudden non power frequency in steady state conditions of voltage.
- ➤ Oscillatory Transients: consists of voltage whose instantaneous value changes polarity rapidly, defined by the spectral content, duration and magnitude. The spectral contents subclasses are high, medium and low frequency. The oscillatory transients with frequency greater than 500 kHz and duration measured in microsecond are high frequency oscillatory transients. The oscillatory transients with frequency 5-500 kHz and duration 10 microseconds are medium frequency oscillatory transients. Back to back energization of capacitor results in oscillatory transient currents in 10 kHz. Transients with frequency components less than 5 kHz and 0.3 to 0.5 milliseconds are low frequency transients. Oscillatory transients with principal frequencies less than 300 Hz can also be found in distribution systems. These are generally associated with ferroresonance and transformer energization.

Principle of overvoltage protection: the main source of the transients are capacitor switching, magnification of capacitor switching transients and lightening. The fundamental principles of overvoltage protection of load equipment are:

- ➤ Limit the voltage across sensitive insulation.
- ➤ Divert the surge current away from the load.
- ➤ Block the surge current from entering the load.
- ➤ Bond ground references together at the equipment.
- ➤ Reduce or prevent surge current from flowing between grounds.
- ➤ Create a low pass filter using blocking or limiting principles.

The surge arresters and transient voltage surge suppressors are widely used. Their main function is to limit the voltage that can appear between two points in the circuit.

10

2.2.2 Long Duration Voltage Variations

- ➢ Overvoltage: this can be defined as an increase in r.m.s voltage greater than 110% at the power frequency for longer than 1 minute. It results from load switching (e.g. switching off large load or energizing large capacitor). It results because the system is too weak for voltage regulation or voltage control is inadequate.

- ➢ Undervoltage: this can be defined as a decrease in r.m.s voltage less than 90% at the power frequency for a duration less than 1 minute. A load switching on or a capacitor bank switching off can cause an undervoltage until the voltage equipment on system can bring voltage back to within tolerances.

- ➢ Sustained interruption: when the voltage has been zero for longer than 1 minute, the long duration voltage variation is considered as a sustained interruption.

Protection against long duration voltage variations: the root cause of most voltage regulation problems is that there is too much impedance in the power system to properly supply the load. Therefore, the load voltage drops too low under the heavy load. Conversely, when the source voltage is boosted to overcome the impedance, there can be an overvoltage condition when the load drops too low. The corrective measures usually involve either compensation for the impedance Z or compensation for the voltage drops IR + jIX caused by the impedance.

The options for improving the voltage regulation are:

- ➢ Add voltage regulators, which boost the apparent VI.
- ➢ Add shunt capacitors to reduce the current I and shift it to be more in phase with the voltage.
- ➢ Add series capacitors to cancel the inductive impedance drop IX.
- ➢ Reconductor lines to change to a larger size to reduce the impedance Z.
- ➢ Change the service transformer to larger size to reduce the impedance Z.
- ➢ Add static var. compensators, which serve the same purpose as capacitors for rapidly changing loads.

There are a variety of voltage regulating devices in use on utility and industrial power systems, and they can be divided into three major classes;

- Tap changing transformers.

> Isolation devices with separate voltage regulators.

> Impedance compensation devices, such as capacitors.

Isolation devices include UPS systems, ferroresant transformers, and M-G sets. These are devices that isolate the load from power source by performing some sort of energy conversion. Therefore, the load side of the device can be separately regulated and can maintain constant voltage regardless of what is occurring at the power supply. The downside of using such devices is that they introduce more losses and may also cause harmonics problems on the power supply system.

Shunt capacitors help maintain the voltage by reducing the current in the lines. To maintain a more constant voltage, the capacitors can be switched in conjunction with the load, usually in small incremental steps to follow the load more closely. Series capacitors are rare because of the extra care in engineering required for the series capacitor installation. The series capacitors compensate for most of the inductance in the system up to the load. If the system is highly inductive, this represents a significant reduction in impedance.

The application of static var. compensators is also possible. These can react within few cycles to maintain a fairly constant voltage by controlling the reactive power production. Such devices are used on arc furnaces and other randomly varying loads where the system is weak.

2.2.3 Short Duration Voltage Variation

- ➢ Interruption: occurs when the supply voltage or load currents decrease to less than 0.1 p.u for a period of time not greater than 1 minute. It results from the power system faults and equipment failure.

- ➢ Dip: this can be defined as a decrease to between 0.1 to 0.9 p.u. in r.m.s voltage from 0.5 cycles to 1 minute. This phenomenon is caused by energization of heavy loads or starting of large motors. A typical voltage dip can be associated with SLG fault on another feeder or substation.

- ➢ Swell: increase between 1.1 and 1.8 p.u. in r.m.s voltage at power frequency for durations from 0.5 cycles to 1 minute. This can be associated with system faults. One way when the voltage swell can occur for temporary voltage rise on unfaulted phase during SLG fault and also can be caused by switching off large load or energizing large capacitor.

Protection against voltage dips and interruptions: solution at higher levels is more costly. The solution close to the load, however, is cheaper. The least cost solution is often for the end user to specify to the supplier that the machine is able to ride through dips of a designated duration and magnitude. Many suppliers can provide the necessary capability if it is specified at the time quotations are requested. At the next higher level, it may be possible to apply UPS system or some other type of power conditioning to the machine control. At higher level "3" some sort of backup power supply with the capability to support the load for a brief period is required. Higher level "4" represents alterations to the utility power system to significantly reduce the number of dips and interruptions.

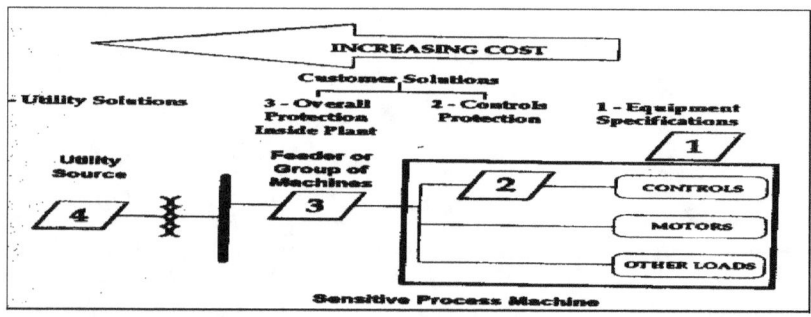

Figure 2-1: Approaches for voltage dip ride through

14

Voltage dips are characterized by three factors; magnitude duration and phase angle jump [14]. Lucid hereunder are the three factors used in characterizing voltage dips:

- **Magnitude:** this is the residual or the remaining voltage during the event. In the case of a three-phase system, the voltage dip can be characterized by the minimum RMS voltage during the dip [20]. If the dip is symmetrical, i.e. equally deep in all three phases that is taken as the magnitude of the voltage dip. If the dip is not symmetrical, i.e. not equally deep in all three phases, the phase with the lowest remaining voltage is used to characterize the event. Magnitude of the voltage dip at a certain point in the system depends mainly on the type and the resistance of the fault, distance to the fault and the system configuration. Calculating the voltage dips requires considering the point of common coupling, and using the voltage divider model [20].

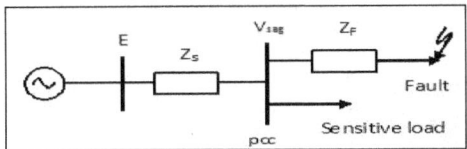

Figure 2-2: Voltage at the point of common coupling during a dip

- **Duration:** the voltage dip duration is mainly determined by the fault clearing time. The amount of time during which the voltage magnitude is below threshold is chosen as 90% of the nominal voltage magnitude. In the case of three-phase systems, the three RMS voltages have to be considered. A voltage dip starts when at least one of the phases of the RMS voltage drops below the threshold, and ends when all three voltages have recorded above the dip threshold [14], [20]

- **Phase Angle Jump:** a short circuit in a power system not only cause a drop in voltage magnitude but also a change in the phase angle of the voltage. In a typical 50Hz system, voltage is a complex quantity which has magnitude and phase angle. A short circuit causes a change in the magnitude, as well as in the phase angle. The phase angle jump can be seen as a shift in zero crossing of the instantaneous voltage. This is not of much concern to most equipment, but may affect power electronics converters using phase angle information for their firing instants [20]

15

2.2.4 Waveform Distortion

- **DC offset:** presence of dc voltage or current in an ac power system. Occur due to result from geometric disturbance or effect of half wave rectification.
- **Harmonics:** sinusoidal voltages and currents having frequencies that are integer multiples of the frequency at which the power system was designed. Harmonic distortion originates from the non-linear characteristics of devices and loads in power systems. All circuits containing both capacitances and inductances have one or more natural frequencies. When one of those frequencies lines up with a frequency that being produced on the power system, resonance can develop in which the voltages and currents at that frequency continue to be at high values.
- **Interharmonics:** the frequency components are not integer multiples of the fundamental frequency. Caused by static converters, cyclo-converters and induction motors.

There are several measures commonly used for indicating the harmonic content of the waveform. One of the most common is total harmonic distortion (THD), which can be calculated for either voltage or current:

$$THD = \frac{\sqrt{\sum_{h=2}^{h_{max}} M_h^2}}{M_1} =$$

THD is a measure of the effective value of the harmonic components of a distorted waveform, that is, the potential heating value of the harmonics relative to the fundamental.

The rms value of the total waveform is not the sum of the individual components, but the square root of the sum of the squares. THD is related to the rms value of the waveform as follows:

$$rms = \sqrt{\sum_{h=1}^{h_{max}} M_h^2} = M_1\sqrt{1+THD^2}$$

All circuits containing both capacitances and inductances have one or more natrual frequencies. When one of those frequencies lines up with a frequency that being produced on the power system, resonance can develop in which the voltages and current at that frequency continue to persist at high values.

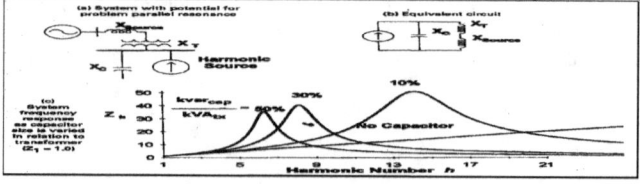

Figure 2-3: Effect of capacitor on the parallel resonant frequency

2.2.5 Principles for Controlling Harmonics

When the harmonics problem occurs, the basic options for controlling harmonics:

1- **Reduce the harmonics currents produced by the load:** For example adding a line reactor in series with PWM drives will significantly reduce harmonic. Transformer connections can be employed to reduce harmonic in three phase systems. Phase shifting half of the six pulse power converters in a plant by 30 degrees can approximate the benefits of 12-pulse loads by dramatically reducing the fifth and seventh harmonics. Delta connected transformers can block the flow of the zero sequence harmonics from the line. Zigzag and grounding transformers can shunt the triples off the line.

2- **Add the filters to either siphon the harmonic currents off the system, block the currents from entering the system, or supply the harmonics currents locally:** The shunt filter works by short circuiting the harmonics currents close to the source of distortion. This keeps the currents out of supply system. Another approach is to apply a series filter that blocks the harmonic currents. This is a parallel tuned circuit that offers high impedance to harmonic currents. One common application is in the neutral of the grounded wye capacitor to block the flow of triplen harmonics while still retaining a good ground at fundamental frequency.

3- **Modifying the system frequency response:** Adverse system responses to harmonics can be modified by a number of methods: adding a shunt filter, adding a reactor to detune the system, changing the size of the capacitor, moving a capacitor to a point on the system with different short impedance or high losses, removing the capacitor and simply accepting the higher losses.

There are two general classes of filters: Passive filters and Active filters

1. **Passive filters:** Passive filters are made of inductance, capacitance and resistance elements. They are relatively inexpensive compared to other means for eliminating harmonic distortion, but they have the disadvantage of potential adverse with the power system. They are employed either to shunt the harmonic currents off the line or block their flow between parts of the system by tuning the elements to create resonance at the selected frequency. The most common type is the notch series filter. An example of common 480 V arrangement is shown in Fig. 3. One important side effect of adding a filter is that it creates sharp parallel resonance with power system at a frequency below the notch frequency. This resonant frequency must be safely away from any significant harmonic.

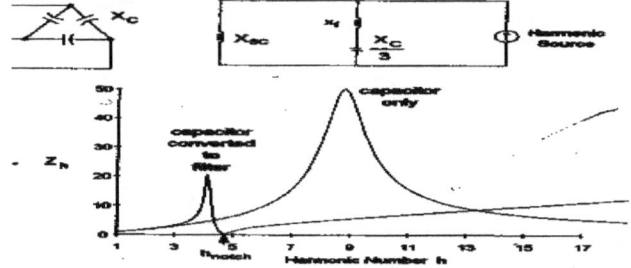

Fig.2.4. Creating a fifth-harmonic notch filter and its effect on system response.

2. **Active filters:** Active filters are based on sophisticated power electronics and are much more expensive than passive filters. However they have distinct advantage that they dont resonate with the system. The basic idea is to replace the portion of the sine wave that is missing in the current in the non-linear load.

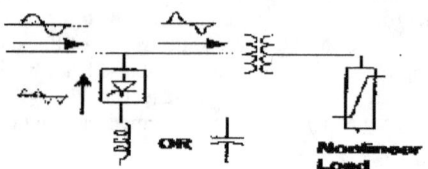

Fig.2.5. Application of an active filter at a load.

2.2.6 Voltage Unbalance

1) **Definition:** Voltage Unbalance can be defined as the maximum deviation from average of 3 phase voltages divided by average of the three phase voltages expressed in percentage. Unbalance can be defined using symmetrical components. The ration of either negative or zero components to the positive equals the percentage unbalance.

2) **Causes:** Voltage unbalance on the network can be caused by amongst other factors;

   - Non-uniform distribution of single phase loads among the three phases of the power system.
   - Unbalanced three phase loads such as arc furnaces.
   - Asymmetrical transmission impedances.
   - Open-wye and open-delta transformer banks.
   - Mismatch in reactive power between generation and demand.

3) **Effects of Unbalance:** Voltage Unbalance being a Power Quality problem itself has some undesirable effects on the network as whole. Listed below are some of the negative effects of unbalance, which need to be taken into consideration at all times;
   - Additional losses results due to higher r.m.s load currents in the supply system.
   - Negative sequence currents will result when asymmetrical voltage is applied and cause unnecessary losses in energy and in performance of rotating loads.
   - Rotating loads will require derating when supplied by asymmetrical voltages.
   - Lifetime of equipment can be reduced due to sustained higher operating temperatures.

4) **Mitigation Measures:** Though it is not within the realm of possibilities to have a system that is completely free from unbalanced voltages, there are some practices already in place to make sure that the levels are kept at the reasonable degrees. Below are four of the popular measures used to mitigate voltage unbalance;
   - Design for balance; transposed transmission lines
   - Reconfiguration/ distribution of loads
   - Formulating best practices for arc furnaces operation
   - Using specific mitigation equipment like compensators.

2.2.7 Stability

As the load demand and the generation change continuously, the system must automatically adjust to the new conditions. Power system stability is the ability to keep the generators in synchronism, and to keep a desired voltage and frequency in the presence of load and generation variations and disturbances.

Power systems are large, complex and highly non-linear systems. Stability analysis has to be performed with simplified models. Depending on the nature of the potential instability, the size of the disturbance, and the time scale, different approaches to modelling and system analysis are used. This leads to a classification of power system stability. This classification is well known to power engineers, and can be found in any book on power system stability. Based on the nature of the potential instability the following classification is made:

1. **Angle stability:** Is the ability to keep the generators in synchronism. This is a problem of balancing active power, as imbalance in mechanical torque and electrical torque makes a generator accelerate or decelerate. If the generator speeds up, the load angle is increased, and the machine automatically takes a larger part of the load. This increases the electric torque and decelerates the machine. If this increase in electric torque is enough to stop the acceleration, the system remains in synchronism.

2. **Voltage stability:** Is the ability to supply the load with a high enough voltage. This is a problem of balancing reactive power. An inductive load supplied via a weak line leads to a large voltage drop across the line. The load voltage will then be low. Since many loads aim to draw constant power, a low voltage implies an increased current, which further increases the voltage drop. If the voltage drop cannot be compensated for by reactive power injection, the result may be a voltage collapse.

3. **Frequency stability:** Is the ability to keep the frequency steady at the reference frequency, for instances 50 or 60 Hz under continuous load variations.

Any disturbance small or large can affect the synchronous operation of the system. For example, there can be sudden increase in load or loss of generation. Another type of disturbance which may occur due to overloading or fault. The stability of the system determines whether the system can settle down to new original steady state
The disturbance can be divided into two categories (a) small or (b) large. A small disturbance is one for which the system dynamics can be analysed from linearised equations. The small changes in load or generation can be termed as small disturbance. However, faults which result in a sudden dip in the bus voltages are large disturbances and require remedial action in the form clearing of the fault. The duration of the fault has a critical influence on system stability. System stability can be divided in the following categories:

1. **Steady state or small signal stability:** A power system is steady state stable for particular steady state operating condition if, following any disturbance, it reaches a steady state operating condition which is close to or identical to the pre-disturbance operating condition.

2. **Transient stability:** A power system is transient stable for a particular steady-state operating condition and for a particular large disturbance or sequence of disturbances if, following that disturbances, it reaches an acceptable steady state condition.

20

2.2.8 Power System Stability Controllers

A cost efficient and satisfactory solution to problem of oscillatory instability is to provide damping for generator rotor oscillations This is conveniently done by providing power system stabilisers (PSS) which are supplementary controllers to the excitation systems. A signal V_s in excitation system is output from PSS, which has input signal from rotor velocity, frequency, electrical power or a combination of these variables. The objective of designing PSS is to provide additional damping torque without affecting the synchronising torque at critical oscillation frequencies.

Also it is well established now that the stability characteristics of the interconnected systems can be improved by power modulation DC links. The control signal used is bus frequency deviations or power flow in the parallel AC tie. With weak AC systems, active and reactive power modulation can be implemented by providing controllers at both rectifier and inverter station. The modulation of extinction angle at the inverter in response to AC voltage signal can avoid voltage instability.

Static Var. Compensator (SVC) connected at the midpoint if a transmission line can help to increase the transfer capability of the line. If auxiliary controllers are provided, utilising signals from locally available measurements, the small signal stability can be improved.

The improvement of transient stability can be achieved not only by adequate system design but also from the use of control action initiated following a disturbance and is temporary in nature. An example is the operation of dynamic breakers using shunt or series connected resistors. Such controllers are termed discrete supplementary controllers. The following controls are listed in IEE.

1. **Dynamic braking:** Mostly connected in shunt which are switched following a fault clearing to correct temporary imbalance between the mechanical power input and the electrical power output of generators.
2. **High speed circuit breaker reclosing**
3. **Independent pole tripping**
4. **Discrete control of excitation system:** The angle signal is generated during the transient disturbance by closing the switch S when there is a sudden drop in terminal voltage followed by the rise in rotor speed above a preset value. The effect of discrete control is to maintain the voltage and consequently the terminal voltage at high level during the positive swing in rotor angle.

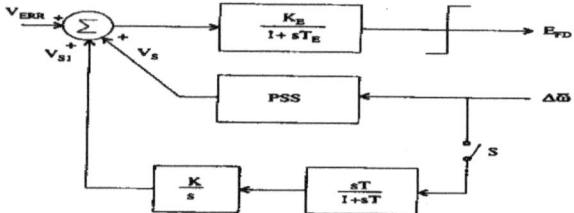

Fig.2 6. Discontinuous excitation system.

5. **Controlled system separation and load shedding:** Load shedding can help to prevent frequency decay and maintain equilibrium between generation and load when there is loss in generation.

6. **Series capacitor insertion:** A series capacitor is inserted by opening a switch at the instant when the fault is cleared. The transient stability limit is higher.
7. **Power modulation of HVDC lines:** Modern HVDC converter stations use thyristor valves for conversion and are controlled to maintain power flow in the line at preset value. As fast control of power flow is achievable due to thyristor controllers, modulation of power is feasible and is implemented. While continuous small signal power modulation is beneficial in improving steady state stability, discrete control of power can result in improving transient stability.
8. **Turbine bypass valving**
9. **Momentary and sustained fast valving:** Fast valving is a means of decreasing turbine mechanical power when a unit is accelerated due to transmission system fault.
10. **Generator tripping:** Selective tripping of generators for transmission line outages can be used to improve stability.
11. **FACTS devices:** With the introduction of SVCs and FACTS devices such as Controllable Series Compensation (CSC) and Thyristor Controlled Phase-Angle Regulators (TCPR), fast control to maintain system security is feasible. FACTS controllers based on high semiconductor devices such as Thyristors can be programmed to provide discrete control action in the event of a major disturbance which may threaten transient stability of the system.

2.2.9 Control of Voltage Instability

The following definitions for voltage stability are defined:

1. **Small-disturbance voltage stability:** A power system at a given operating state is small disturbance voltage stables if, following any disturbance, voltage near loads is identical or close to pre-disturbance values.

2. **Voltage stability:** A power system at a given operating state and subjected to a given disturbance is voltage stable if voltages near loads approach post disturbance equilibrium values.

3. **Voltage collapse:** Following voltage instability, a power system undergoes voltage collapse if the post-disturbance equilibrium voltages near loads are below acceptable limits. Voltage collapse is total or partial blackout.

The incidence of voltage instability increases as the system is operated close to its maximum loadability limit. Environmental and economic constraints have limited the transmission network expansion, while forcing the generators to be sited far a way from the load centres. This resulted in stressing the existing transmission network.

The reactive power compensation close to the load centres as well as the critical buses in the network is essential for overcoming voltage instability. The location, size and speed of control have to be selected properly to have maximum benefits. The SVC and STATCON provide fast control and help improve stability.

The design of suitable measures in the event of voltage instability is also necessary. The application of undervoltage load shedding, controlled system separation and adaptive or intelligent control are steps in this direction.

2.2.10 Subsynchronous Resonance

Subsynchronous resonance is an electric power condition where power electric network exchanges energy with turbine generator at one or more of the natural frequencies of the combined system below the synchronous frequency of the system. There are two categories of SSR problems. These are

1-Self-Excitation: Subsynchronous frequency currents entering the generator terminals produce subsynchronous frequency terminal voltage components. These voltage components may sustain the currents to produce the effect that is termed as self-excitation. There are two types of self-excitation, one involving only rotor electrical dynamics and the other involving both rotor electrical and mechanical dynamics. The first one is termed as induction generator effect while the second one is called as torsional interaction.

a) **Induction Generator Effect:** As the rotating mmf produced by the subsynchronous frequency armature currents is moving slower than the speed of the rotor, the resistance of the rotor (at that subsynchronous frequency) viewed from the armature terminals is negative. When the magnitude of this negative resistance exceeds the sum of the armature and network resistances at a resonant frequency, there will be self-excitation

b) **Torsional Interaction:** Generator rotor oscillations at a torsional mode frequency f_m induce armature voltage components at frequencies (f_{em}) given by $f_{em} = f_o \pm f_m$. When f_{em} is close to f_{er}, the subsynchronous torques produced by the subsynchronous voltage component sustained. This interplay between the electrical and mechanical systems is termed as torsional interaction.

2- Transient Torque: System disturbances resulting from switching in the network can excite oscillatory torques on the generator rotor. Due to SSR phenomenon, the subsynchronous frequency components of torque can have large amplitudes immediately following the disturbance, although they may decay eventually. Each occurrence of these high amplitude transient torques can result in expenditure of the shaft life due to fatigue damage.

2.2.11 Countermeasure for Subsynchronous Resonance

Some of the countermeasures as denoted by [2]

1. **Static blocking filter**: This is inserted in series with the generator step-up transformer winding on the neutral end of the transformer high voltage winding. SBF is three-phase filter made up of separate filters connected in series. Each section of the filter is a high Q parallel resonant circuit tuned to block electric currents at frequencies corresponding to each torsional mode.

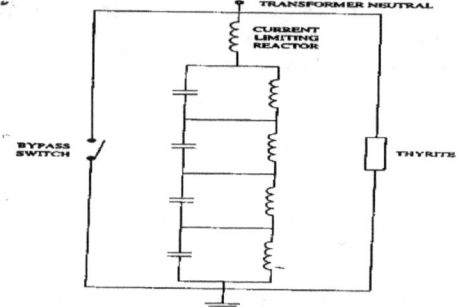

Fig. 2.7. Static blocking filter.

2. **Bypass damping filter:** Damping resistor connected in series with parallel combination of a reactor and capacitor, which is tuned at system frequency. Thus the filter has a very high impedance at the system frequency. The damping resistor is effective at subsynchronous frequencies.

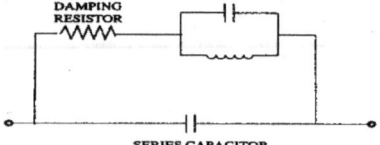

Fig. 2.8. Bypass damping filter.

3. **Supplementary excitation damping control (SEDC):** The control objectives are met by parallel processing of each mode. Each mode is isolated from the relevant speed signal measurement, by bandpass filtering at each torsional frequency. The necessary phase and gain compensation is provided individually for each component mode and the output of SEDC is obtained by summing the control signal for all modes.

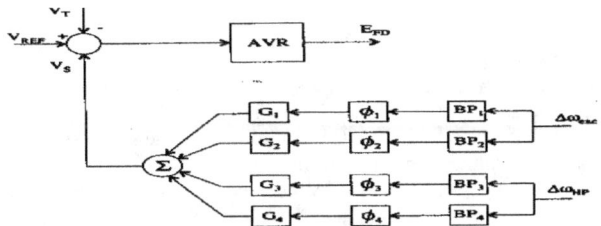

Fig. 2.9. SEDC model control structure.

4. **NGH damping:** If a sinusoidal voltage of frequency f_o is combined with a DC voltage, it is seen that, for the combined voltage, some half cycles are longer than the normal half cycle period of $1/2f_o$. The voltage across the series capacitor is a combination of the fundamental frequency, DC and subsynchronous frequency components. The basic principle of the NGH damping scheme is to dissipate capacitor charges whenever the measured half cycle period exceeds the nominal. This is done by inserting resistor across the capacitor through thyristor switches. The thyristor stops conducting whenever the capacitor voltage reaches zero. Thereafter, the measurement of half cycle period restarts from new voltage zeros. No thyristor fires for half cycle, which are shorter than the set point.

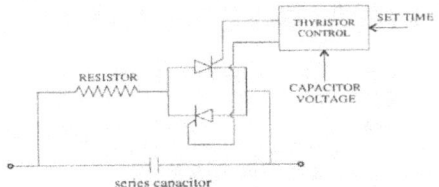

Fig.2.10. Basic NGH -SSR damps scheme.

5. **Dynamic stabiliser:** The shunt subsceptance of TCR is given by $B(\sigma) = (\sigma - \sin\sigma)/(\pi X_L), \rho = 2(\pi - \alpha)$. The firing angle is modulated around quiscent point in response to the oscillation of generator rotor. In the absence of the rotor oscillations, the dynamic stabiliser appears as continuous reactive load. The design of dynamic stabiliser is such that it generates sufficient current of appropriate phase to compensate for the critical subsynchronous frequency currents in the generator armature due to the network resonance.

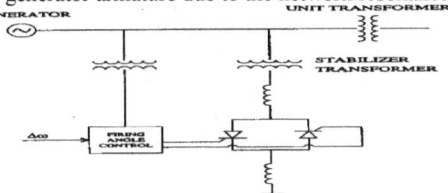

Fig. 2.11. Dynamic stabiliser.

27

2.3 Harmonics Simulations

This section on harmonic resonance, the Matlab simulation online diagram is presented and further explained. The operation of the "Three-Phase Harmonic Filters", is explained by means of an example of an HVDC installation, where the AC harmonic shunt filters are used to:

1. Reduce harmonic voltages and currents in the power system.
2. Supply the reactive power consumed by the network.

To illustrate the above concepts, a 1000MW (500kV, 2kA) HVDC rectifier is simulated. The HVDC rectifier is built up from two 6-pulse thyristor bridges connected in series. The converter is connected to the system with a 1200MVA "Three-Phase Transformer" with three windings. A 1000MW resistive load is connected to the DC side through a 0.5H smoothing reactor. The filters set is made up of the following components;

- One capacitor bank (C1) of 100MVar modelled by a three-phase series RLC load.
- One C-type high pass filter tuned to the 3rd of 150MVar
- One double-tuned filter 11/ 13th of 150MVar
- One high pass filter tuned to the 24th of 150MVar

Total MVar rating of the filters set is then 600MVar. A three phase circuit breaker is used to connect the filters set to the AC bus.

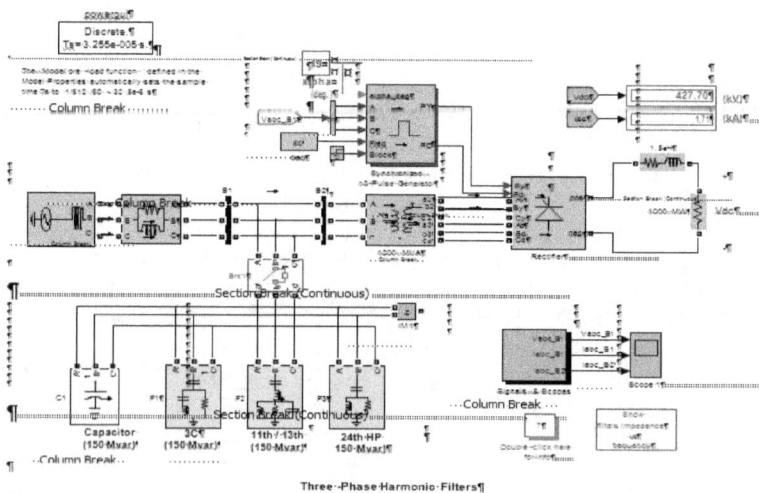

Figure 2-12: Matlab Simulation Online Diagram for Three Phase Harmonic Filter

28

Three conditions are considered and they are illustrated below.

1-Condition 1: No capacitor and no filter connected

No capacitor and no filter are connected, that means the breaker is off. The simulation diagram is shown in Figure 2.11. The circuit is simulated with the circuit breaker open, meaning the power system has no capacitor bank, nor harmonic filters set. Waveforms of three different cases are presented in the figures that follow. In this condition we do expect to see some harmonics on the system. The currents and voltages waveforms are distorted meaning the system is infested with harmonics. The Fast Fourier Transform (FFT) analysis showing the spectrum of the voltage shows the distorted signal, where it is shown that the Total Harmonic Distortion (THD) reaches the heights of 17.78%.

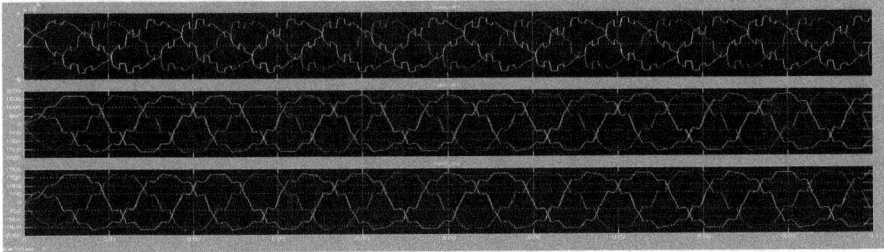

Figure 2-13: Voltage and currents waveforms from the scope

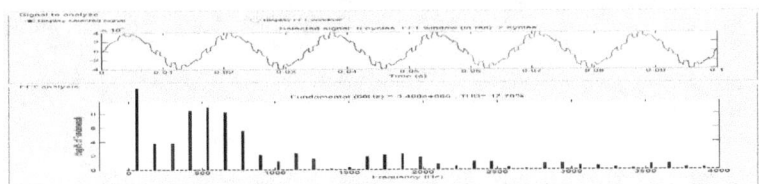

Figure 2-14: Spectrum of the Voltage from FFT Analysis

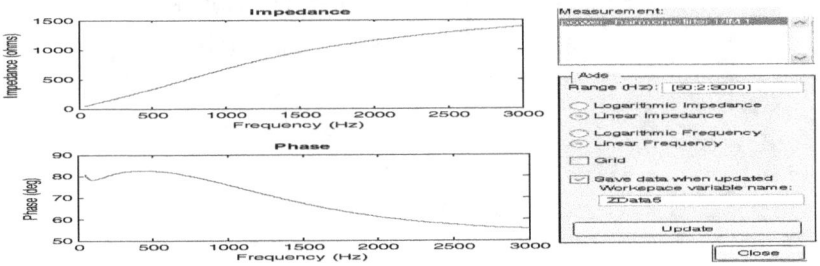

Figure 2-15: Frequency Scan

29

2- Condition 2: Capacitor and no filter

Capacitor connected with no filter, this means that the breaker is on and the filter is removed. Waveforms of this conditions from the simulation are shown in the figures to follow. The waveforms of the current and the voltage show a high level of distortion because the harmonics are not filtered. The THD in this case rises from 17.78% to 23.00%. The fundamental frequency of this system is 60Hz. Also the system shows some parallel resonance, and the resonant frequency is around 400Hz, with peak magnitude input impedance = 2000Ω. So if there is any harmonic at this frequency it can be amplified and cause distortion on the system. For example, if 1A of 400Hz harmonic is injected at the point of measuring the impedance, it will cause distorted voltage of 2000V of this harmonic. To mitigate harmonics we use filters, and a system with filters is simulated in the third condition

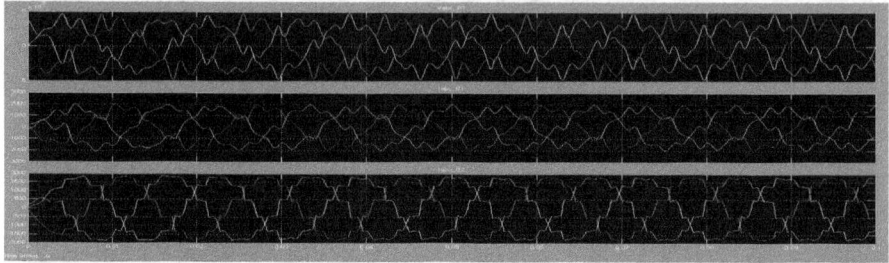

Figure 2-16: Voltage and current waveforms from scope

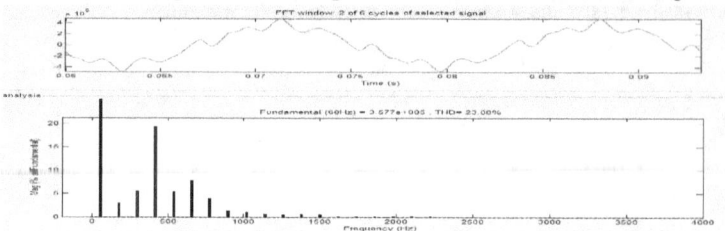

Figure 2-17: Spectrum of the voltage from FFT analysis

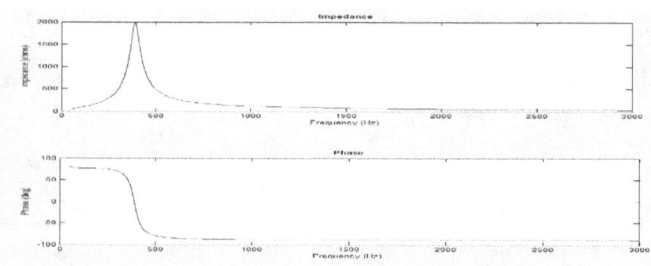

Figure 2-18: Frequency Response

30

3-Condition 3: Capacitor and filters set connected

Capacitor and filter, that is to consider the system with the breaker on and the filters set connected to the system. The harmonic filters reduce the harmonics generated by the converter to about 0.7%. From the plot of impedance against frequency of the system with harmonic filters there are four resonant frequencies

- 3rd harmonic frequency: $3*60 = 180Hz$
- 11th harmonic frequency: $11*60 = 660Hz$
- 13th harmonic frequency: $13*60 = 780Hz$
- 24th harmonic frequency: $24*60 = 1440Hz$

So the three harmonic filters will shunt the 3rd, 11th, 13th, and 24th harmonics from the line. But the filter and capacitor will cause shunt resonance at frequencies 180Hz, 300Hz, 670Hz, and 800Hz. If there is any harmonic at these frequencies it can be amplified and cause distortion on the system. The shunt resonance at frequency 800Hz has peak magnitude of input impedance = 400Ω. For example, if 1A of 800Hz harmonic is injected at the point of measuring the impedance, it will cause distorted voltage of 400V of this harmonic.

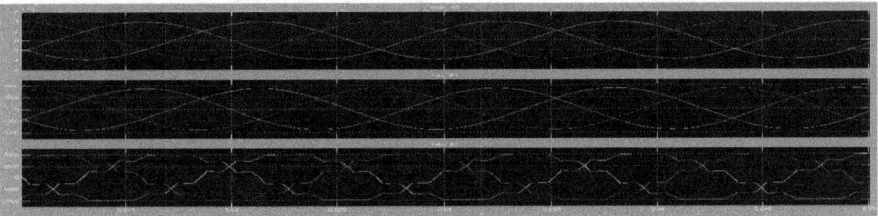

Figure 2-19: Voltage and frequency waveform from scope

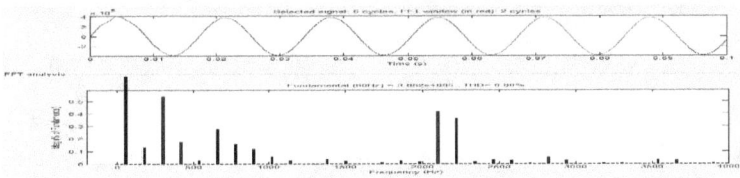

Figure 2-120: Voltage spectrum from FFT analysis

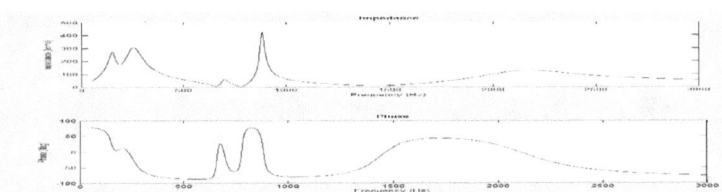

Figure 2-21: Frequency Response

31

CHAPTER 3- POWER QUALITY STANDARDS

3.1 Introduction

Power Quality standards are required to ensure that power utilities deliver an ideally perfect Power Quality, and that the customers receive the quality of power they need. The purpose of Power Quality standards is to protect the utility and customer equipment from failing when the voltage, current, or frequency deviates from normal. These standards provide this protection by setting measurable limits as to how far the voltage, current or frequency can deviate from normal. By setting these limits, Power Quality standards help utilities and their customers gain agreements as to what acceptable and unacceptable levels of service are. So, clearly, a knowledge and an understanding of the relevant standards by the electricity utility is essential in effectively managing Power Quality on the supply network [2].

Organizations playing a major role in the developing and authorization of Power Quality standards include the Institute of Electrical and Electronic Engineers (IEEE) and International Electro technical Commission (IEC). Other organizations that are active in the research, development and analysis of Power Quality standards are amongst others; American National Standards Institute (ANSI), Electric Power Research Institute (EPRI), National Electric Code (NEC), Information Technology Industrial Council (ITIC) and National Electrical Manufactures Association (NEMA) [2]. A brief list of typical IEEE and IEC standards is available in Appendices.

3.2 Swaziland/ South African Standards

In 2007, the South African energy regulator formulated the NRS-2: 2007 document which serves as a framework for power utilities as they go about selling power to customers. The document is entitled, *"Electricity Supply – Quality of Supply. Voltage Characteristics, Compatibility Levels, Limits and Assessment Methods"*. In Swaziland, we have our very own version titled as the South African version; DS/SZNS 027: 2012, and it was published in 2012. These documents are technically the same document; specify compatibility levels, voltage characteristics and assessment methods, which can be used by utilities, customers, and the regulator in managing the level s of Power Quality supplied at the point of supply to customers.

The NRS 048 takes recommendations from international (IEC and Cigre), European (CENELEC), and North American (IEEE) standards and reports, as well as reports and data available locally. The SZNS takes recommendations from the NRS 048.
In these standards voltage quality parameters that might affect the normal operation of the electricity dependent processes of customers are discussed. Each of the voltage quality parameters is described and where appropriate, compatibility levels, limits and assessment methods are specified. Compatibility levels and limits provide measures of acceptable voltage quality at the point of supply to end customers of electricity utilities. Assessment methods define how measured values are assessed over a given time. The assessed values are compared with the compatibility levels or limits.
Below are the Power Quality parameters taken into consideration in both these documents; NRS 048 and SZNS 027:
- Frequency
- Voltage Dips
- Voltage Flicker
- Voltage Harmonics and Interharmonics
- Voltage Swells
- Voltage Transients and Surges
- Voltage Unbalance
- Voltage Regulation
- Mains Signalling

From the standards document SZNS 027, typical compatibility levels are shown for a number of power quality parameters are shown.
- **Frequency:** The standard frequency shall be 50Hz. The level for the national grid is ±2%, that is ±1Hz, and for island network is ±2.5%, that is ±1.25Hz.
- **Voltage Regulation:** For customers supplied at low voltage (LV: < 1kV), the standard voltage shall be $\sqrt{3}$ x230V phase to phase, and 230V phase to neutral.For voltages less than 500V, the levels are ±10% and for voltages greater than 500V, the levels are ±5%.
- **Voltage Unbalance:** For Low Voltage, Medium Voltage (MV: 1kV < V_n < 33KV), and High Voltage (HV: 33kV < V_n < 220kV) networks the voltage unbalance shall not exceed 2%, and for Extra High Voltage (EHV: 220kV < V_n < 400kV) the voltage unbalance shall not exceed 1.5%.
- **Voltage Harmonics and Harmonics:** The THD for LV and MV networks shall not exceed 8%, and for HV and EHV it shall not exceed 4%.

33

CHAPTER 4 POWER QUALITY MONITORING

4.1 Introduction

To minimize the Power Quality disturbances and to devise suitable corrective and preventive measures, efficient detection and classification techniques are required in the emerging power systems. Typical standards that utilities conform to in as far as Power Quality monitoring is concerned are amongst others; EN50160, which is a European standard, IEEE 1159: Recommended Practices for Power Quality Monitoring, for North America and some Asian countries, as well as NRS 048, a South African standard [2].

Solving Power Quality (Power Quality) problems not only depends on the technology applied to solve the problem, but a profound knowledge of the Power Quality phenomena, the applied solution and the electrical installation is needed to find the most effective solution. Troubleshooting and simple fixing measures are short term solutions; knowledge is the only way to find long term solutions. Monitoring is an essential analyzing tool used to improve the availability of power, and it requires investment in equipment, time, education and labor.
There are two types of Power Quality surveys; reactive and proactive. The reactive type of monitoring is initiated in response to customer complaints. This type of monitoring concentrates on the specific customer or point of common coupling. Its objective is to identify the nature and the source of the problem. The proactive type encompasses continuous monitoring, and is motivated by a number of reasons; establishing conformance with standards, gaining a better understanding of system performance, to inform customers of what Power Quality levels can be expected on the network amongst other reasons [1].

This study will focus on the latter type of Power Quality monitoring, that is the continuous type of monitoring, for the purposes of benchmarking the system performance of SEC's distribution network, that is, to perform a gap analysis of the network.
Of the many important reasons to monitor Power Quality, the primary reason which stands over all others is economic. Effects on equipment and process operations can include malfunction, damage, process interruptions and many other uncalled for conditions. Such interruptions are costly since a profit-based operation is interrupted unexpectedly and must be restored to continue production. Equipment damage and subsequent repair cost both money and time.

The objectives of the monitoring program need to be clearly defined before-hand because they determine the choice of measuring equipment, method of data collection, selection of disturbance threshold, data analysis requirements and the overall level of the required effort. The purpose of the survey will also influence the method of reporting the results to best suit the intended audience. Fundamental questions that the Power Quality monitoring planner needs to answer are; where to measure, what to measure and how to measure. Clearly defining the objectives of the monitoring program will automatically answer the above questions, and hence the whole monitoring program will be made simpler.

Typical monitoring objectives;

34

- To characterize system performance
- To characterize specific problems
- Enhanced power quality service
- Predictive or just-in-time maintenance

Where to measure: the whole process of Power Quality monitoring comes with some considerable costs; costs of instruments, costs associated with data communication and data storage, and the time required to process and analyze the data. The above mentioned factors make Power Quality monitoring a rather expensive exercise. By virtue of necessity, the utility needs to limit the number of sample sites for the Power Quality survey. A methodology is required for selecting the best sites the best sites for installing monitors. The best sites also are determined by the objectives of the survey. Listed here under are some of the factors that can be considered during the selection of the best sites.

- Characterizing the Power Quality levels being experienced by customers: the actual customer service entrance location is ideal, because it includes the effect of step-down transformers supplying the customer. This location can also characterize the customer load current variations and harmonic distortion levels.
- Characterizing Power Quality levels on electric utility distribution feeders: monitoring locations should be on the actual feeder circuits.
- At LV or MV side: LV monitoring shows the levels seen by domestic customers. MV monitoring shows the disturbance levels seen by larger industrial and commercial customers. As a result, MV monitoring appears to be the better choice over LV monitoring, presumably because one monitoring site can cover many a customer. However, some Power Quality disturbances emanate in the LV system and will not be seen at their worst at the MV side. This calls for going back to the objectives of the survey and determine exactly what questions the survey outcomes will answer. A possible compromise would be to monitor the MV system at the substation and at selected customer entry points.
- Number and locations of individual monitors: cost constraints will be the major consideration, and possibly limitation. It is necessary to determine the best possible monitoring locations so as to obtain representative measurements. Indication of the overall Power Quality performance can be obtained provided the monitoring sites are selected without partiality [2], [3].

What to measure: Power Quality encompasses a wide variety of conditions on the distribution network. The wide range of conditions that must be characterized during the survey presents a challenge in both the requirements of the monitoring equipment and the data collection process. This will also be to a larger extent made easy by clearly defining the principal objectives of the survey.

- *Disturbances that have the most impact on customers*: customers are mostly affected by voltage dips, so the survey must mainly focus on voltage dips. Voltage variations and armonics are of significant concern to customers.
- *Capabilities of the monitoring equipment:* this will also influence the choice of which Power Quality parameters can be measured. Some instruments do not have the capability to measure voltage fluctuations, while others may not have sufficiently fast sampling rates to accurately capture high frequency disturbances such as impulsive transients.
- *Surveys designed to evaluate conformance to standards:* each and every Power Quality

disturbance; dips, flicker, unbalance, harmonics and frequency variation has to conform to set standards, so for example a survey designed to evaluate conformance with harmonic standards may only require steady-state monitoring of harmonic levels.

- *Specific industrial problems:* this might only encompass the monitoring of RMS voltage variations such as voltage dips.*Benchmarking system performance:* this kind of factor requires a reasonably complete monitoring effort. The complete types of disturbances have to be monitored; transients, dips/swells, interruptions, undervoltages/overvoltages, harmonic distortions and voltage fluctuations [2].

How to measure: the type of instruments used will determine what can be measured and the degree of accuracy, as well as the survey period. We have numerous reasons for undertaking continuous monitoring; enhanced customer service, system benchmarking, and real-time alarming on Power Quality events. Continuous monitoring should normally span for 1-2 years. This is appropriate for system benchmarking, general study of the network and for providing disturbance level information to customers.

The major limitations of continuous monitoring; require significant investment in data communication and storage infrastructure and imposed additional work load on those responsible for analyzing and reporting the data [2].

4.2 The VectoGraph Recorder

This recorder is designed to measure and record voltage quality related parameters. Capable of measuring voltage dips/ swells, harmonics, voltage magnitude, voltage unbalance, amongst others. The VectoGraph being the instrument used by SEC in its monitoring program is written about below. The ProvoGraph and ImpedoGraph and Impedo Duo are briefly explained in the Appendices section.

1. Hardware Interfaces
2. Power Supply

The VectoGraph can be powered from 90 – 300Vac, 50 Hz, as well as from 50 – 150Vdc. A constant voltage transformer (CVT) is used to step-up the supplied AC voltage, and the secondary voltage is rectified and applied to high voltage capacitors. The high voltage capacitors are used to store the energy required to ride through dips and interruptions. A DC/DC converter is used to supply power to the instrument, with an electronic fuse that protects the CVT from overloading. When powered from DC this recorder consumes 3 Watts [9].

4.2.1 Communications Port

This is an RS232 serial port, used to communicate with a PC/laptop through a standard serial cable at 115200 baud. As seen in Figure 4.0 below, there is another port, RJ45 interface modem line, for connecting a modem, which facilitates a remote communication with a PC/laptop.

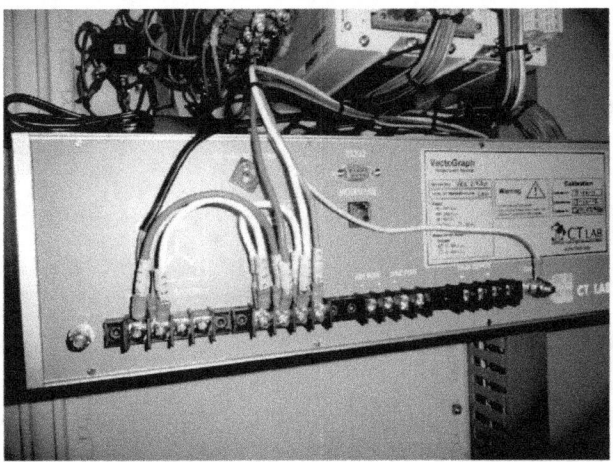

Figure 22: The rear view of the VectoGraph

4.2.2 Calibration

The VectoGraph is software calibrated, and the calibration software is stored on-board the instrument and is unique to that instrumentRecording Memory

This recorder has a fairly large memory storage capacity, that is, about 220 000 memory locations. The First In, First Out concept is used in the event the memory gets used up; old data is automatically deleted to give way to "fresh" data. [9]

4.2.3 Access Control

The VectoGraph has 4 access levels which are password protected to make sure that some information can be only accessed by authorized personnel.
1. **View**: connect, and view meters and configurations
2. **Retrieve:** connect, view meters and configurations, and retrieve recordings.
3. **Configure:** access the Retrieve level and can synchronize clocks, change configuration and configure clock synchronization.
4. **Administration:** can perform the functions on the previous three levels, as well as change passwords and upgrade firmware [9].

4.3 Power Quality Recorder Manager

Each one of the recorders mentioned above is accompanied by software that facilitates the storage, and processing of the data collected by the recorder. The software is called the Power Quality Recorder Manager (PQRM), and is developed by the same company that manufactures the recorders.

The CT Lab family of recorders; VectoGraph, ImpedoGraph and ProvoGraph are stand-alone 3- phase Power Quality recorders recording Power Quality events, trends and statistics. Recordings are communicated to a database from where it can be viewed, printed and exported in different ways. The PQRM organizes recordings and support single roaming instruments and networks of remotely installed instruments.

The PQRM software can be installed on any number of PCs, from the Internet via http://www.ctlab.com/pqrm or from CD. Tools that are included in the software are amongst others; Configuration Template Editor, Address Book, Online Tasks, Retrieve Recordings, Event Browser, Trend Viewer, Preferences, Adjust PC Date and Time and Synchronize the Recorder Clock [8].

4.4 SEC's Power Quality Monitoring Program

The Swaziland Electricity Company is currently on a Power Quality monitoring program, which came into full operation in January 2014. Taken verbatim from a tender document by the company titled "Quality of Supply Measurement Services";

The Swaziland Electricity Company (referred to as SEC hereafter) has expressed its desire to measure and ascertain the quality of supply at different point in its network. This is per the dictates of the national standard SZNS 027 for „Quality of Supply" as well as the SEC Customer
Service Charter. Moreover, this is now a Southern African Power Pool (SAPP) requirement on all interconnections in the Southern African Power Grid. SEC"s gains from such an exercise would be data that would eventually help the company identify elements that contribute to undesirable events in its network and thereafter implement measures to mitigate them.

The VectoGraph was selected as the best instrument to be used on the Power Quality monitoring program that is on-going at SEC. Seven VectoGraphs

Date installed	Cellphone number	Serial Number	Feeder	Voltage (kV)	Substation
21/01/2014	78477430	VEC2930	2010	132	Stonehenge
20/01/2014	78477457	VEC2931	3590	66	Hhelehhele
21/01/2014	78477466	VEC2926	6040	132	kaLanga
20/01/2014	78477472	VEC2929	4010	132	Nhlangano2
21/01/2014	78477462	VEC2927	1470	66	Simunye
21/01/2014	78477433	VEC1441	7060	400	Dwaleni2 Camden
21/01/2014	78747469	VEC1443	7030	400	Dwaleni2 Maputo

Table 4.1: Information on the location of the Power Quality monitoring instruments

4.4.1 Getting Started with the VectoGraph

For the PQ Monitoring program, SEC chose the VectoGraph to carry out the function of collecting the PQ data on the network. The VectoGraph is therefore further discussed on the paragraphs below.

With the PQRM already installed on the PC, a serial connection to the VectoGraph from the software on the PC needs to be made as follows;

- Connect the serial cable to the PC port of the VectoGraph and an unused port of the PC.
- Start the PQRM software application.
- Select the correct serial port in the *Connect to a VectoGraph* panel and then click on *Connect*.
- The window will change to display the status of the connection and after a while the window will change to indicate that a connection is established.

4.4.2 Single Phase Connection

- The power has to be switched off
- The voltage input range written on the instrument should correspond to the voltage to be measured at the particular location.
- Earth the instrument to a safe earthing point.
- Link the voltage measurement N input to the auxiliary N input and connect to system neutral.
- All the voltage measurement inputs as well as the auxiliary should be linked together and then to system live.
- Select the correct voltage input range: 50 – 150Vrms on 120V range; 150 – 300Vrms on 240V range (50/60Hz).

4.4.3. Three phase delta connection (3-wire)

- Power has to be switched off and the VT panel fuses should be removed.
- Voltage input range on the instrument label should correspond to the voltage to be measured.
- The VT should be connected to a 3-wire system.
- Connect voltage measurement N input to the VT phase that is earthed.
- Connect voltage measurement N input to the VT phase that is earthed.
- Connect the three voltage measurement inputs to the remaining VT outputs
- Select voltage selector switch for correct voltage input range: 50 – 150Vrms on 120V range; 150 – 300Vrms on 240 range (50/60Hz).
- Switch on VT supply and then auxiliary supply.

CHAPTER 5- RESULTS AND DISCUSSION OF RESULTS

5.1 Introduction

This chapter presents the data as collected from the installed VectoGraphs from three out of the six substations considered in the program. The substations were selected as per their nominal operating voltage. In the program we have six substations with installed VectoGraphs, which are measuring and recording time-stamped Power Quality events as they occur on the network. The substations making up the program are;
1. Edwaleni2_7030_400kV
2. Edwaleni2_7060_400kV
3. Nhlangano2_4010_132kV
4. Stonehenge_2010_132kV
5. KaLanga_6040_132kV
6. Simunye_1470_66kV
7. Hhelehhele_3590_66kV

These substations form virtually the whole of the transmission network of SEC. Here three voltage levels are represented; 400kV, 132kV and 66kV, and they were strategically selected to give a true overview of the Power Quality set-up of the whole network.

A typical electricity network conveys energy from multiple sources (generating stations) to multiple loads. The entire system including generators, loads, and everything in between is a single integrated and dynamic system- any change of voltage, current, impedance, etc. at one point instantaneously brings about a change at every other point on the system. This is meant to illustrate that these substations are able to give a true picture of what is actually happening on the entire network.

For the purposes of this paper the number of substations of interest was further cut down to three based primarily on the three different voltage levels. The substations under scrutiny here are:
1. Edwaleni2_7060_400kV
2. KaLanga_6040_132kVkV
3. Simunye_1470_66kV

Three Power Quality disturbances are considered for each station; Voltage Dips, Harmonics, and Voltage Unbalance, for a period of three months from January to March. The order of the disturbances will be the Voltage Dips, followed by the Voltage Harmonics (IHD and THD), and lastly the Voltage Unbalance.

The data is interpreted and analyzed after the presentation of the results. Power Quality theory from literature and Power Quality Standards are invoked for an in-depth analysis of the data. This analysis also helps in bridging the gap between the literature part and the practical aspect. Some of the questions answered by the analysis are amongst others;
- What do the values tell us?
- What were we expecting to see?
- Is it normal as compared with other utilities?

- Is the disturbance from upstream or downstream?
- What are the possible causes of the results ?
- Should SEC invest on improving the performance?
- Is there anything that SEC can do to mitigate the disturbances?

5.2 Voltage Dips:

Data from the VectoGraphs installed on the three substations for voltage dips is tabulated below, for the three months. The voltage dips as recorded by the VectoGraphs are time-stamped, so the date and time at which each dip occurs are shown in the tables. Also on the tables, the duration; which is the time taken by the voltage to fall below the dip threshold and rise up again to the nominal voltage is shown, as well as the residual voltage; which is the voltage that remains during the dip, expressed as a percentage of the nominal voltage. Tables 5.1 to 5.8 are shown below to show also the dip characterization.

5.2.1 Voltage Dips for Simunye_1470_66kV for three months

				January
Class	Residual Voltage (%)	Duration(s)	Time	Date
Y	84.36	0.07	19:08:16	27-01-2014
Y	83.72	0.51	08:30:57	28-01-2014
Y	85.00	0.51	06:26:56	29-01-2014
Y	83.61	0.07	13:41:16	30-01-3014
Y	85.65	0.34	15:29:29	31:01:2014

Table 5.1: Simunye_1470_66kV January dips

				February
Class	Residual Voltage (%)	Duration(s)	Time	Date
Y	81.60	0.08	14:21:42	04/02/2014
Y	85.22	0.39	22:00:52	05/02/2014
Y	81.39	0.03	02:38:23	10/02/2014
Y	80.52	0.03	14:04:15	12/02/2014
Y	81.69	0.03	13:49:33	20/02/2014
Y	82.78	0.03	18:22:19	25/02/2014

Table 5-2: Simunye_1470_66kV February dips

44

	March			
Class	Residual Voltage (%)	Duration(s)	Time	Date
Y	86.14	0.04	17:04:48	03/03/2014
Y	70.58	0.10	00:24:10	04/03/2014
Y	72.84	0.06	04:20:11	05/03/2014
Y	86.07	0.08	21:14:54	06/03/2014
X1	62.55	0.10	12:02:17	08/03/2014
Y	72.71	0.27	13:23:59	09/03/2014

Table 5.2: Simunye_1470_66kV March dips

5.2.2 Voltage Dips for KaLanga_6040_132kV for three months

				January
Class	Residual Voltage (%)	Duration(s)	Time	Date
X1	65.87	0.07	12:07:09	25/01/2014
Y	81.89	0.59	11:09:39	26/01/2014
Y	83.77	0.03	18:27:37	27/01/2014
Z1	81.45	0.75	10:29:05	28/01/2014
Y	86.32	0.36	8:25:04	29/01/2014
Y	74.60	0.12	15:08:48	30/01/2014
Y	81.63	0.49	17:27:39	31/01/2014

KaLanga_6040_132kV January dips

				February
Class	Residual Voltage (%)	Duration(s)	Time	Date
S	69.54	0.48	18:08:15	4/2/2014
S	59.9	0.57	15:51:49	5/2/2014
Y	89.34	0.40	0:35:14	6/2/2014
Y	85.98	1.13	10:44:30	11/2/2014
Y	74.04	0.08	5:15:49	12/2/2014
Y	89.63	0.13	6:30:20	13/2/2014
Y	81.3	0.39	23:06:51	17/2/2014
Y	88.48	0.15	19:08:09	19/2/2014
Y	82.58	0.06	15:29:52	20/2/2014
Z2	67.3	2.99	4:46:45	21//2/2-14

KaLanga_6040_132kV February dips

	March			
Class	Residual Voltage (%)	Duration(s)	Time	Date
Y	77.69	0.06	14:17:16	17/03/2014
X1	68.63	0.08	15:13:04	17/03/2014
Y	89.07	0.09	15:13:04	17/03/2014
Y	89.74	0.03	15:13:04	17/03/2014
Y	89.85	0.03	15:13:04	17/03/2014

KaLanga_6040_132kV March dips

5.2.3 Discussion and Analysis of the voltage dips

Voltage Dips are defined by their magnitude and their duration. The magnitude in our case is expressed as a percentage of the nominal voltage, where the r.m.s value of the voltage is taken during the dip; and this is termed the residual voltage, and it can either be 10%, 90% or anything in between. Magnitudes beyond that do not fit the description of a voltage dip. The duration of the voltage dip is nothing but the time taken by the voltage to drop below the dip threshold up to the time it recovers. By definition it ranges from 20ms to 3 s.

The SZNS 027 standard, alongside the NRS 048 of South Africa outlines a dip characterization criterion based on the residual voltage and the duration. the shows the dip characterization.

Duration			Residual Range of
$0.6 < t < 3$ s	$150 < t < 600$ ms	$20 < t < 150$ ms	Voltage (%)
Y	Y	Y	$> U_r > 85$ 90
Z1			$> U_r > 80$ 85
	S		$> U_r > 70$ 80
Z2		X1	$> U_r > 60$ 70
		X2	$> U_r > 40$ 60
		T	$> U_r > 0$ 40

Dip characterization according to the Swaziland standard

The Y class is dominating in this characterization, covers a much wider spectrum. The residual voltage range is from 70 to 90% and the duration spans the whole range of a typical voltage dip. The S class is characterized by a residual voltage range of 40 to 80% with a duration range of 150 to 600ms. The Z1 and Z2 classes have residual voltage ranges of 70 to 85% and 0 to 70%, respectively, duration range is 0.6 to 3s for both classes. Last in the characterization is the T class, which has a residual voltage range and a duration range of 0 to 40% and 20 to 600ms, respectively. Dip characterization by both standards is based on the philosophy that;

- Utilities should manage protection performance times – some dips may be allowed to occur more frequently than others.
- Utilities should place a particular emphasis on managing the number of faults that occur close to a particular customer.
- Customers should specify the dip sensitivity of their process equipment to enable appropriate mitigation measures to be considered, so as to limit the number of fault events that actually affect the plant.

SEC network is highly infested with Y class dips; 20ms to 3s duration and a residual voltage range of 70 to 90%. These class of dips have a distinct characteristic of requiring minimum plant compatibility, meaning it does not require the supplier nor the customer to press the panic button. The Y dips can occur more frequently but they are not much of a threat to customer equipment.

The frequency at which the dips occur increase as we move downstream; Simunye_1470_66kV recorded 591 (sampled by taking daily averages on this paper) dips in one month, while dips recorded at Edwaleni2_7060_400kV hardly pass the number 10. We have every reason to believe that motors starting at the mill can be taken as the major cause of the voltage dips on that part of the network. Some of the methods employed to mitiate voltage dips on industrial installations include; tap changing transformer, saturable reactor regulators, constant voltage transformer, phase controlled,   amongst   others.

5.3 Voltage Harmonics

Data for voltage harmonic distortion is presented in the nine tables below. In the subject of Power Quality we are mainly concerned with the quality of the voltage. In harmonic distortion a current distortion in the load side causes a voltage distortion which is what is seen on the supply network. Three stations are considered for the data analysis for the period of three months from January to March.

Literature states that harmonics are categorized into even harmonics, odd harmonics and interharmonics. In our discussion we look at a sample of odd harmonics which are said to be a threat to the system. The individual harmonic distortions are considered as opposed to the total harmonic distortions. So the harmonic number states the sequence number of the harmonic; 3, 5, 7, 9, 11, 13 and 23, for the 3rd, 5th, 7th, 9th, 11th, 13th, 21st and 23rd harmonics, respectively.

When referenced to the 50Hz fundamental frequency of the network, these harmonic components have frequencies of 150Hz, 250Hz, 350Hz, 450Hz, 550Hz, 650Hz, 1050Hz and 1150Hz, respectively. These are presented against their individual compliance level, which is the threshold which the individual harmonic is expected not to pass.

The assessed value is the percentage contribution of the individual harmonic distortion to the overall or total harmonic distortion, as measured on the network. Also the compliancy status is indicated, considering the assessed value against the compliance level. The compliance level is the value stated in the standard document, and is used as the threshold.

As stated in the literature review section; suppose the nominal voltage of the system is 415V, then the 3rd, 5th, and 7th components have amplitudes of 13V, 15V and 17V, respectively. We therefore have the above expressed in percentage of the nominal voltage on an individual basis.

Below are the tables showing the individual voltage harmonic distortions for the three substations over the period of three months.

5.3.1 Individual Voltage Harmonic Distortion compliance for Simunye_1470_66kV
for three months

Compliance Status	Value Assessed (%)	Level Compliance (%)	Harmonic Number
January			
Comply	0.22	5	3
Comply	2.05	6	5
Comply	1.08	5	7
Comply	0.22	1.5	9
Comply	0.32	3.5	11
Comply	0.22	3	13
Comply	0.00	0.3	21
Comply	0.00	1.41	23

Simunye_1470_66kV IHD compliance for January

Compliance Status	Value Assessed (%)	Level Compliance (%)	Harmonic Number
February			
Comply	0.22	5	3
Comply	2.59	6	5
Comply	1.08	5	7
Comply	0.22	1.5	9
Comply	0.32	3.5	11
Comply	0.22	3	13
Comply	0.00	0.3	21
Comply	0.00	1.41	23

Simunye_1470_66kV IHD compliance for February

March			
Compliance Status	Value Assessed (%)	Level Compliance (%)	Harmonic Number
Comply	0.22	5	3
Comply	3.67	6	5
Comply	2.59	5	7
Comply	0.43	1.5	9
Comply	0.54	3.5	11
Comply	0.32	3	13
Comply	0.00	0.3	21
Comply	0.00	1.41	23

Simunye_1470_66kV IHD compliance for March

5.3.2 Individual Voltage Harmonic Distortion compliance for KaLanga_6040_132kV for three months

| | | | January |
Compliance Status	Level Assessed (%)	Level Compliance (%)	Harmonic Number
Comply	0.33	5	3
comply	2.21	6	5
Comply	1.55	5	7
Comply	0.33	1.5	9
Comply	0.55	3.5	11
Comply	0.66	3	13
Comply	0.00	0.3	21
Comply	0.00	1.41	23

KaLanga_6040_132kV IHD compliance for January

| | | | February |
Compliance Status	Value Assessed (%)	Level Compliance (%)	Harmonic Number
Comply	0.33	5	3
Comply	2.87	6	5
Comply	1.66	5	7
Comply	0.33	1.5	9
Comply	0.66	3.5	11
Comply	0.88	3	13
Comply	0.00	0.3	21
Comply	0.00	1.41	23

| March | | | |
Compliance Status	Assessed Value (%)	Compliance Level (%)	Harmonic Number
Comply	0.33	5	3
Comply	4.20	6	5
Comply	2.54	5	7
Comply	0.55	1.5	9
Comply	0.77	3.5	11
Comply	0.88	3	13
Comply	0.00	0.3	21
Comply	0.00	1.41	23

KaLanga_6040_132kV IVHD compliance for March

5.3.3 Analysis and Discussion of Individual Voltage Harmonic Distortion

Harmonic voltage distortion is caused by the flow of harmonic current through system impedance. Non-linear loads in the network draw currents with waveforms which are not similar to waveform of the supply voltage. These harmonic current flows back into the supply network, together with the impedance of the network bring about the voltage harmonics.

The data above is for the individual harmonic distortion (IHD) compatibility levels. The IHD indicates the contribution of each harmonic frequency to the distorted waveform. Here the odd harmonics are considered. The compatibility levels as per the Power Quality standards document of Swaziland- SZNS 027: 2012 for individual harmonics are shown in the. Here both the odd and even harmonics are shown; with the odd harmonics further categorized into multiples of triplens and non-triplens harmonics, for multiples of 3 and non-multiples of three, respectively, the magnitude of each harmonic is expressed as a percentage of the declared voltage.

Even Harmonics		Odd Harmonics			
		Triplens		Non-Triplens	
Magnitude (%)	Harmonic order h	Magnitude (%)	Harmonic order h	Magnitude (%)	Harmonic order h
2	2	5	3	6	5
1	4	1.5	9	3	7
0.5	6	0.5	15	3.5	11
0.5	8	0.3	21	3	13

Individual Harmonic Distortion compatibility levels

Data from the monitoring program includes only the odd harmonics; 3, 5, 7, 9, 11, 13, 21, and 23; that is 3 triplens and 5 non-triplens. If we look at the assessed values of each of the individual harmonics we can see that they are way below the compatibility levels, making the system to be in full compliance with the standards. This is supposed to tell us that the network is free from harmful harmonics, hence we do not expect equipment malfunction emanating from harmonics.

The instruments as they are placed on the transmission network, we do expect lesser level of harmonics, because here we are far from the non-linear loads, which are prevalent on the customer side. We should expect a considerable level of harmonics since the customer are directly connected on the distribution network. This is evident on the data as shown above; on the 66kV substation the level of the harmonics is considerable when compared with that on the 400kV substation.

The IHD, as stated in the previous paragraphs, indicates the contribution of each harmonic frequency to the distorted waveform. The net deviation of the waveform due to all the harmonics is called Total Harmonic Distortion (THD). The data collected also shows the THD for each substation for the duration of the three months.

Data for the THD for the three substations is shown in the table , where the compatibility level, assessed value and compliancy status for each month is shown. The compliance level is supposed to be the limit set by the standards, and if passed that will mean that station is non- compliant and mitigating measures need to be formulated and implemented

		Compliance		Assessed Level		Compatibility Level	
Mar	Feb	Jan	Mar	Feb	Jan		Substation
Comply	Comply	Comply	1.56	1.56	1.41	4	Edwaleni2
Does Not Comply	Comply	Comply	4.81	3.39	2.69	4	KaLanga
Does Not Comply	Comply	Comply	4.38	2.83	2.40	4	Simunye

THD compliance in the three substations

For LV and MV networks the THD of the supply voltage shall not exceed 8%, on the other hand for HV and EHV networks the THD of the supply voltage shall not exceed 4%. This is stated in the document of the Power Quality standard used in Swaziland.

As seen on the table above, the data shows some compliance, which shows that on the transmission network the problem of harmonics is not so prevalent. On the month of March, Simunye and KaLanga proved to be non-compliant. These substations are closer to the customer, hence we may expect the levels of the harmonics to be considerable.

These two substations are showing us that downstream there is a prevalent of non-linear loads which are a major cause of harmonics on the supply voltage, and as the harmonics are formed on the customer end, they propagate upstream causing a stir in the entire network.

The generic solution for solving or mitigating harmonics on the network is the use of harmonic filters. Capacitors combined with inductors can be used to limit the effects of harmonics. Harmonics are a bad Power Quality issue to happen on the network, so it is of best interest that it should be prevented at all costs.

5.4 Voltage Unbalance

In a sinusoidally balanced supply system, the three line-neutral voltages are equal in magnitude, and are phase displaced from each other by 120 degrees. Any difference that exists in the three voltage magnitudes and/or shift in the separation from 120 degrees is said to give rise to an unbalanced supply. This is the condition in a poly phase system, three-phase in most cases, in which the r.m.s voltage of the line-neutral voltages and/ or the phase angles between line voltages are not equal.

Presented below is the data as collected in real-time from the three substations for the period of three months.

5.4.1 Voltage Unbalance for Simunye_1470_66kV for three months

January			
Average	95 % Highest	Maximum	Date
0.69	0.80	0.90	27/01
0.68	0.80	0.90	28/01
0.67	0.80	1.00	29/01
0.65	0.80	1.00	30/01
0.65	0.80	1.00	31/01

Simunye_1470_66kV voltage unbalance data for January

February			
Average	95% Highest	Maximum	Date
0.63	0.80	1.00	1-Feb
0.64	0.80	1.00	2-Feb
0.66	0.90	1.00	3-Feb
0.68	0.90	1.10	4-Feb
0.69	0.90	1.10	5-Feb
0.70	0.90	1.10	6-Feb
0.70	0.90	1.10	7-Feb
0.70	0.90	1.10	8-Feb
0.69	0.90	1.10	9-Feb
0.66	0.80	1.10	10-Feb
0.64	0.80	0.90	11-Feb
0.62	0.70	0.80	12-Feb
0.63	0.70	0.80	13-Feb
0.63	0.70	0.80	14-Feb
0.63	0.70	0.80	15-Feb
0.62	0.70	0.80	16-Feb
0.60	0.70	0.80	17-Feb
0.61	0.80	0.90	18-Feb
0.61	0.80	0.90	19-Feb
0.59	0.80	0.90	20-Feb
0.58	0.80	0.90	21-Feb
0.58	0.80	0.90	22-Feb
0.60	0.80	0.90	23-Feb
0.62	0.80	0.90	24-Feb
0.60	0.70	0.80	25-Feb
0.59	0.70	0.80	26-Feb

Simunye_1470_66kV voltage unbalance data for February

53

March			
Average	95% Highest	Maximum	Date
1.05	1.30	1.40	1-Mar
1.08	1.30	1.40	2-Mar
1.06	1.30	1.40	3-Mar
1.04	1.30	1.40	4-Mar
1.03	1.30	1.40	5-Mar
0.99	1.20	1.40	6-Mar
0.98	1.20	1.40	7-Mar
0.95	1.20	1.40	8-Mar
0.89	1.20	1.40	9-Mar
0.86	1.20	1.40	10-Mar
0.84	1.20	1.40	11-Mar
0.81	1.20	1.40	12-Mar
0.78	1.20	1.40	13-Mar
0.74	1.00	1.20	14-Mar
0.74	1.00	1.10	15-Mar
0.78	1.00	1.10	16-Mar
0.79	1.00	1.10	17-Mar
0.80	1.00	1.10	18-Mar
0.80	1.00	1.10	19-Mar

Simunye_1470_66kV voltage unbalance data for March

5.4.2 Voltage Unbalance for KaLanga_6040_132kV for three months

January			
Average	95% Highest	Maximum	Date
1.40	1.90	2.10	22-Jan
1.39	1.80	2.10	23-Jan
1.37	1.80	2.10	24-Jan
1.40	1.80	2.10	25-Jan
1.33	1.80	2.10	26-Jan
1.28	1.80	2.10	27-Jan
1.22	1.70	2.10	28-Jan
1.20	1.60	1.90	29-Jan
1.18	1.60	1.90	30-Jan
1.15	1.60	1.90	31-Jan

KaLanga_6040_132kV voltage unbalance data for January

February

Average	95% Highest	Maximum	Date
1.07	1.40	1.90	1-Feb
1.07	1.40	1.90	2-Feb
1.08	1.40	1.90	3-Feb
1.08	1.40	1.90	4-Feb
2.07	1.40	1.60	5-Feb
1.04	1.40	1.60	6-Feb
1.01	1.30	1.60	7-Feb
1.01	1.30	1.60	8-Feb
1.01	1.30	1.80	9-Feb
1.01	1.30	1.80	10-Feb
1.00	1.30	1.90	11-Feb
0.94	1.30	1.90	12-Feb
0.94	1.30	1.90	13-Feb
0.93	1.30	1.90	14-Feb
0.91	1.30	1.90	15-Feb
0.87	1.20	1.90	16-Feb
0.85	1.00	1.90	17-Feb
0.84	1.00	1.20	18-Feb
0.87	1.00	1.20	19-Feb
0.86	1.00	1.20	20-Feb
0.87	1.10	1.20	21-Feb
0.89	1.10	1.20	22-Feb
0.91	1.10	1.20	23-Feb
0.91	1.10	1.20	24-Feb
0.91	1.10	1.20	25-Feb
0.91	1.10	1.20	26-Feb
0.91	1.10	1.20	27-Feb
0.91	1.10	1.20	28-Feb

Table 5-24: KaLanga_6040_132kV voltage unbalance data for February

March			
Average	95% Highest	Maximum	Date
0.92	1.10	1.20	1-Mar
0.93	1.10	1.20	2-Mar
0.96	1.10	1.30	15-Mar
0.96	1.10	1.30	16-Mar
0.92	1.10	1.30	17-Mar
0.93	1.10	1.30	18-Mar
0.88	1.10	1.30	19-Mar

Table 5-25: KaLanga_6040_132kV voltage unbalance data for March

5.4.3 Analysis and Discussion of Individual Voltage Unbalance:

The standard stipulates that the threshold for voltage unbalance in LV, MV and HV networks should be 2%, foe EHV networks the value drops to 1.5%. Data collected from the substations as shown above reveals th network is not bad when it comes to voltage unbalance. This parameter can be entirely controlled by the su making if easies for the utility to put in place preventive and corrective measures. The only thing that is ex of SEC is to make sure that the status quo is maintained, because we cannot guarantee a perfectly ideal sit where we have a perfectly balanced supply withzero unbalance. This will mean doing everything possible to sure that the mandatory 1.5% limit is not exceeded. Some of the methods that can be done to achieve th making sure that the transmission lines are transposed, re-configuration or correct distribution of loads, as employing the use of compensators.

5.5 Conclusion:

The network as it stands on the transmission part shows a high level of compliance, and if this results are anything to go by the network is conveniently good.

The shortcoming of our results is that we do not have instruments further downstream where poor Power Quality levels are expected to be high. The instruments are limited to the transmission level of the network. Since we have stated that most Power Quality problems are prevalent on the customer premises, as the equipment on the customer premises is the primary cause for Power Quality problems, it would be advisable to install the instruments also on the <11kV network, which is the distribution network.

In some of the substations we have more missing data as compared to available data, making our data to be less valid for investigational purposes. This can be blamed on the network used to remotely connect the instruments with the analyzing stations. The data we have though is data enough to make informed decisions in as far as the Power Quality status of the network is concerned.

The objective of the project on data collection was successfully achieved, as data was collected through the use of VectoGraphs installed on the transmission network in six substations. The access to the data through the website developed by CT Lab called Power Quality portal also made it possible for me to achieve this objective.

Data from Power Quality portal is very informative, and the events are time-stamped. For example, in the case of a voltage dip the duration, residual voltage, channels affected, dip class, and the time it occurred are recorded making life easier for the user.

CHAPTER 6- CONCLUSION, RECOMMENDATIONS and FURTHER WORK

6.1 Conclusion

The study had three main objectives as stated in the introductory chapter.
- Studying the subject of Power Quality; disturbances, monitoring, as well as standards.
- Monitoring the power disturbances on the SEC network.
- Comparing the findings with literature.

Power Quality has been studied from written and online literature, and from experts on the power fi
Swaziland. A number on Power Quality standards documents were consulted. These include loca
international standards. Though there are some variations in these standards, it is recommendable that i
choose to follow a certain standard you have to maintain consistency. Fortunately in Swaziland we have the
027 which is a local standard, making it easy for the power utility to follow one set of standards.

With the Power Quality instrument installed in six substations across the network, monitoring the netwo.
Power Quality related issues was made possible, and since the instrument is permanently installed
substations, monitoring and measurements are being done in real- time day after day, minute after minut
second after second. Data is being stored in the memory of the instrument waiting to be retrieved at the oper
will.

Comparing the outcome and literature was to merely verify that the practical aspect is in line with what lite
says, though power is a dynamic phenomenon, research has to be done on a continual basis so as to be a
with the changes that come with the times.

The hypothesis was to prove or to disapprove that an understanding of Power Quality as a subject, an
monitoring of the network was the first step towards understanding the status of the network, and to
preventive and corrective measures for the network.

6.2 Recommendations

To achieve the objective of providing a good quality of supply, SEC needs to from time to time liaise with customers, so as to get their feedback on the power supplied.

Continuously monitoring the network for Power Quality related issues is a step in the right direction, because the operation of the network will be understood, and all the events on the network will be based on real-time. Putting adequate resources on the quality of supply exercise is highly recommendable. This entails having the personnel tasked with handling this task do it to perfection, make sure that data collected is presented as meaningful information that will help in taking the necessary actions on the network.

Expanding the network of VectoGraphs to distribution level where it is believed that most Power Quality issues are experienced, and in that way there will be some assurance that the network is covered in its entirety.

REFERENCES

[1] Baggini, A. (2001). Handbook of Power Quality. University of Bergamo, Italy.

[2] Sankaran, C. (1998). Power Quality. CRC Press

[3] Herath, C. (2008). Power Quality data management and reporting methodologies. MSc Thesis, University of Wollongong.

[4] Nicholson, G. (2006). Investigation of data reporting techniques and analysis of continuous Power Quality data in the Vector distribution network. MSc Thesis, University of Wollongong.

[5] Patne N.R. (2008). Factors affecting characteristics of Voltage Sag Due to Fault in the Power System. Serbian Journal of Electrical Engineering. Vol.5, No.1, May 2008, 171-182

[6] Lidong, Z. and Bollen, H. (2000). Characteristics of voltage dips (sags) in power systems. IEEE transactions on power delivery, Vol. 15, No 2, April 2002.

[7] The McGraw Hill Company (2004). Electrical Power Quality McGranaghan. Available: http://www.digitalengineeringlibrary.com.

[8] The McGraw Hill Company (2004). Power Quality primer. Available: http://www.digitalengineeringlibrary.com.

[9] CT Lab. (2004). Power Quality recorder manager user's guide. Version 3.2.0. Stellenbosch, 7600, South Africa.

[10] NRS 048-2: 2007. Electricity Supply- Quality of Supply; Voltage Characteristics, Compatibility Levels, Limits and Assessment Methods.

[11] SZNS 027: 2012. Electricity Supply- Quality of Supply; Voltage Characteristics, Compatibility Levels, Limits and Assessment Methods.

[12] CT Lab. (2005). Quality of supply- ProvoGraph and VectoGraph user's guide. Version 2.0. Stellenbosch, 7600, South Africa.

[13] CT Lab. (2012). Impedo Duo brochure. Stellenbosch, 7600, South Africa.

[14] Bachry, A. (2006). Power Quality in distribution systems involving spectral decomposition. PhD Dissertation. University of Trier.

[15] Maharajan, D. (2012). Power Quality monitoring. SRM University.

[16] Almeida, A. Moreira, L and Delgado, T. Power Quality Problems and New Solutions.

ISR- Department of Electrical and Computer Engineering. University of Colombia.

[17] Kamble, S. and Thorat, C. Characteristic Analysis of Voltage Sags in Distribution Systems using RMS Voltage Method. ACEEE International Journal on Electrical and Power Engineering, Vol. 03. No. 01. February 2012

[18] Moussa, A. El-Gammall, M. and Abdallah, E. A New Methodology for on-line Power Quality Assessment. Department of Electrical Engineering, Alexandria University.

[19] Dwivedi, U. Developing a Smart Power Quality Monitoring System: CI Application, Raji Gandhi Institute of Petroleum Technology, Raebareli, India

[20] Mandangombe, T. Integration of Wind Energy System into the Grid: Power Quality and Technical Requirements. MSc Thesis. Department of Electrical Engineering, University of Cape Town. March 2010

[21] Kazunovic, M etal Power Quality Assessment Using Advanced Modelling, Simulation and Data Processing Tools. Texas A&M University.

[22] Slyvaklakis, E. Automating Power Quality. Department of Electric Power Engineering and Department of Signals and Systems, Chalmers University of Technology. PhD Thesis. May 2002

APENDIX A: STANDARDS

A.1: IEEE Standards

There are a number of PQ standards developed by IEEE, amongst them are:
- IEEE Standard 1159-1995, Recommended Practices for Monitoring Electric Power Quality. This standard defines various PQ terms and categorizes IEEE standards by the various PQ standards.
- IEEE Standard 1409-2012, IEEE Guide for Application of Power Electronics for Power Quality.
- IEEE Standard 1159.3-2003, IEEE Recommended Practices for the Transfer of Power Quality Data

A2: IEC Standards

IEC refers to PQ standards as Electromagnetic Compatibility (EMC) standards. This illustrates that IEC is mainly concerned about the compatibility of end-user equipment with the utility's electrical supply system. The IEC 61000 series standards manage EMC for each PQ disturbance type by defining a boundary value known as the compatibility level. For a given disturbance type, the compatibility level is in between the emission level and the immunity level. The compatibility level is chosen such that compatibility is achieved for most (95%) equipment most (95%) of the time.

The IEC Standard 61000.2.2 is cited as the most conventional standard in the 61000 series. It specifies the compatibility levels of the following PQ disturbances:
- Voltage fluctuation and flicker
- Harmonics up to and including order 50
- Interharmonics up to the 50th harmonic
- Voltage distortions at higher frequencies (above the 50th harmonic)
- Voltage dips and short supply interruptions
- Voltage unbalance
- Transient overvoltages
- Power frequency variation
- D.C components
- Mains signaling levels

APENDIX B: POWER QUALITY RECORDERS

B.1: The ProvoGraph Recorder:

This recorder is designed to comply with NRS048 Class 2, and is capable of recording voltage dips, voltage swells and voltage trends. Voltage unbalance is automatically extracted from the voltage trend information. The ProvoGraph is housed in a small aluminium extrusion. Auxiliary supply and the voltage inputs are available on screw terminals [9].

The instrument calculates the true RMS value of each input phase once every ½ cycle. The sampling rate is 32 samples per cycle. Two thresholds bind the RMS voltage upper and lower limits. If any one of the voltages exceeds the threshold, an event is triggered. The captured event contains the following information;
- Date and time of occurrence
- Duration of event for each phase
- Averaged RMS voltages just before the event
- Minimum and maximum ½ cycle RMS value of each value during the event.

The ½ cycle RMS readings are averaged over a 10-minute interval to obtain a voltage trend. An internal lithium battery is used to power the non-volatile Random Access Memory (RAM) and the on-board real time clock during the absence of mains power. A separate rechargeable battery is used to sustain the instrument for up to 5 seconds during long voltage dips or short outages.

The ProvoGraph is field upgradeable, and new firmware can therefore be downloaded to the instrument if required. The instrument and its included software support modem and serial port communication. An optional three level password scheme protects the ProvoGraph from unauthorized access [9].

B.2: The ImpedoGraph Recorder

This is a three phase power quality recorder- a stand-alone instrument designed to record PQ anomalies for weeks or even months at a time. The anomalies are stored in on-board non-volatile memory. Supply interruptions longer than 10 seconds interrupt recording, but no recordings are lost and recording continues as supply returns.

All external connections to the ImpedoGraph are connected to the terminals shown on the figure below.

There is a power light emitting diode (LED) on the front panel indicating the presence of mains supply. Valid mains are 80 to 150Vrms and valid frequency is from 40 to 70Hz. The instrument consumes 45VA. During voltage dips and short interruptions an internal high voltage capacitor bank supplies power to the ImpedoGraph for up to 10 seconds. If the interruption is longer than 10 seconds, the instrument will shut itself down without any loss of recordings.

The instrument has 4 differential voltage inputs, and the inputs are available on an eight terminal barrier strip connector on the front of the instrument. Each of the four voltage inputs has a built- in high-speed peak detector that captures positive and negative polarity fast voltage transient events up to 2kV.

The ImpedoGraph has 4 galvanically insulated current inputs. Current transducers provide galvanic isolation, excellent AC and DC response, very low input impedance and a high fault level withstand capability.

Eight high impedance differential voltage inputs are used to capture external status events.

- *PC port*: RS232C serial port used for communication with a PC/laptop through a standard serial cable at 115200 baud

- *Modem port*: used to interface to external modems or any other modem like devices.

- *Ethernet Port*: 10BaseT UTP connector for connecting to Ethernet networks. IP address, subnet mask, and gateway address can be set via software application.

- *GPS port*: RS232C serial port dedicated to time synchronization.

- *SCADA interface port*: RS232C serial port with a built-in RS232 to RS485 converter.

The ImpedoGraph complies with the IEC 61000-4-30 Class A standard, with voltage and current waveforms with very high accuracy and 16-bit resolution. All the analog inputs are sampled simultaneously at a sampling rate of 128 samples per cycle with 4 times oversampling to reduce background noise levels. The instrument measures voltage dips, voltage swells, poor voltage regulation, voltage transients, under frequency and current inrush.

www.ingramcontent.com/pod-product-compliance
Lightning Source LLC
Chambersburg PA
CBHW072015230526
45468CB00021B/1601